SPECTROPIA

SPECTROPIA;

OR,

SURPRISING SPECTRAL ILLUSIONS.

SHOWING

GHOSTS EVERYWHERE,

AND OF ANY COLOUR.

WITH SIXTEEN ILLUSTRATIONS.

APPLEWOOD BOOKS

Carlisle, Mass.

978-1-4290-9693-5

1 2 3 4 5 6 7 8 9 10

PRINTED IN CHINA

John Filmer,
Engraver and Printer.

INTRODUCTION.

THE following Illusions are founded on two well-known facts; namely, the persistency of impressions, and the production of complementary colours on the retina.

The explanations are divided into two parts. The first consists of directions for seeing the spectres. The SECOND, a brief and popular, as well as a scientific, description of the manner in which the spectres are produced, and is intended for the use of those who may wish to know more of this subject than is contained in the first Part.

As an apology for the apparent disregard of taste and fine art in the plates, such figures are selected as best serve the purpose for which they are intended.

DIRECTIONS.

To see the spectres, it is only necessary to look *steadily* at the dot, or asterisk, which is to be found on each of the plates, for about a quarter of a minute, or while counting about twenty, the plate being well illuminated by either artificial or day light. Then turning the eyes to the ceiling, the wall, the sky, or better still to a white sheet hung on the wall of a darkened room (not totally dark), and looking rather steadily at any one point, the spectre will soon begin to make its appearance, increasing in intensity, and then gradually vanishing, to reappear and again vanish; it will continue to do so several times in succession, each reappearance being fainter than the one preceding. Winking the eyes, or passing a finger rapidly to and fro before them, will frequently hasten the appearance of the spectre, especially if the plate has been strongly illuminated.

Those who use gaslight will find it convenient, after having looked at the plate as above described, to extemporise a darkened room by having the gaslight turned low.

The size of the spectres will be determined by the distance of the observers from the plate, and from the surface against which they are seen ; being larger the nearer the plate, and smaller the nearer the surface ; so that short-sighted persons will see them larger than long-sighted, if both are equidistant from the surface against which they are seen.

Should any one not be able to see the spectre's features, the reason will be, either that the eyes have been allowed to wander, or the head to move, while looking at the plate.

Many persons will see some one coloured spectre better than the others, in consequence of their eyes not being equally sensitive to all colours.

The colours in the plate will be found to reverse themselves in the spectres, as explained elsewhere, the spectres always appearing of the complementary colour to that of the plate from which it is obtained. Thus, blue will appear orange, and orange blue, &c.

LIST OF THE PLATES.

———•◦•———

PLATE I.

This winged figure of Victory will give a white spectre by artificial light (rather green by daylight), the red wreaths green, the green roses red, and the orange stars blue.

PLATE II.

This black figure will give a white spectre.

PLATE III.

This will give a dark spectre.

PLATE IV.

This green figure will give a red spectre.

PLATE V.

And this red figure will give a green spectre.

PLATE VI.

This orange figure will give a blue spectre.

PLATE VII.

And this blue figure an orange spectre.

PLATE VIII.

This purple hand will give a yellow spectre.

PLATE IX.

And this yellow figure of Victory will give a purple spectre with green wreath.

PLATE X.

The face of this figure will come out green in the spectre, the garment red, and the cloud white.

PLATE XI.

This black skeleton will make a white spectre.

PLATE XII.

This skeleton will also give a white spectre, with a yellow mantle.

PLATE XIII.

This figure and broom will give a violet spectre with red cloak and hat and white moon.

PLATE XIV.

These figures will give green and white spectres.

PLATE XV.

This Cupid will give a rose-coloured spectre, with bow and arrow white.

PLATE XVI.

This is a rainbow with colours reversed, the spectre of which will be found a good resemblance of nature, especially when seen on a cloudy sky.

———•◦•———

The colours of the spectres produced by these figures will not only be subject to a slight variation in different eyes, but also by the degree in which the plates are illuminated while being looked at.

POPULAR AND SCIENTIFIC DESCRIPTION.

———————•◦•———————

It is a curious fact that, in this age of scientific research, the absurd follies of spiritualism should find an increase of supporters; but mental epidemics seem at certain seasons to affect our minds, and one of the oldest of these moral afflictions—witchcraft—is once more prevalent in this nineteenth century, under the contemptible forms of spirit-rapping and table-turning. The modern professor of these impostures, like his predecessors in all such disreputable arts, is bent only on raising the contents of the pockets of the most gullible portion of humanity, and not the spirits of the departed, over which, as he well knows, notwithstanding his profane assumption, he can have no power.

One thing we hope in some measure to further in the following pages, is the extinction of the superstitious belief that apparitions are actual spirits, by showing some of the many ways in which our senses may be deceived, and that, in fact, no so-called ghost has ever appeared, without its being referable either to mental or physiological deception, or, in those instances where several persons have seen a spectre at the same time, to natural objects, as in the case mentioned by Dr. Abercrombie, in his work on "The Intellectual Powers:"—"A whole ship's company were thrown into the utmost consternation, by the apparition of a cook who had died a few days before. He was distinctly seen walking ahead of the ship, with a peculiar gait, by which he was distinguished when alive, from having one of his legs shorter than the other. On steering the ship toward the object, it was found to be a piece of floating wreck."

A ghost, according to the general descriptions of those who fancy they have been favoured with a sight of one, appears to be of a pale phosphorescent white, or bluish white colour; usually indistinct, and so transparent that objects are easily seen through it. When moving, it glides in a peculiar manner, tho legs not being necessary to its locomotion.

All the senses are more or less subject to deception, but the eye is pre-eminently so; especially in the case of individuals who are in ill health, because the sensibility of the retina is then generally much exalted, as is also the imagination.

We may divide the illusions to which the sense of sight is liable into four kinds. First, mental, or those arising in the brain itself, and only referred to the eye. Second, those produced by the structure of the eye. Third, those arising from the impressions of outward objects on the retina. Fourth, those produced by various combinations of the foregoing. It is only the second and third we shall have occasion to touch upon. But before we can well understand their nature, it will be necessary to get a slight knowledge of the structure of the eye, and some idea respecting the nature of light.

With perhaps the exception of the ear, the eye is the most wonderful example of the infinite skill of the Creator. A more exquisite piece of mechanism it is impossible for the human mind to conceive. The annexed diagram (Fig. 1) of a horizontal section of this organ will give a better idea of its general structure than whole pages of letterpress. It will be seen to consist of a globe of three envelopes or coats, which are kept distended by three transparent humours or lenses: the aqueous (*e*), the crystalline (*f*), and the vitreous (*g*). The outer coat (*a*) is dense, white, and fibrous. In front of the eye it gives place to a perfectly transparent one, called the cornea (*d*). The next coat, the choroid (*b*), is vascular, very black on its internal surface, in order that light falling on it through the pupil (*h*) may not be reflected. The pupil is an opening through a diaphragm which is called the iris (*i*), from its colour varying in different individuals. It has the power of expanding and contracting the pupil, for the purpose of regulating the supply of light to the retina (*c*), or third and last coat which lies immediately on the choroid. It is transparent, very complex, and the only part of the eye we shall carefully consider. The following diagram (Fig. 2) represents a section of it magnified 250 diameters. *a* is called the limitary membrane, and forms its innermost surface, or that which is next the vitreous humour; *b* consists of the layer of optic nerve fibres; *c* is a layer of grey nerve cells; *d*, two layers in which the principal retinal blood vessels are spread out; *e*, two layers of granular matter; *f*, Jacob's membrane, or layer of rods and cones. Fig. 3 will give some idea of the supposed connexion between these various parts, the same letters referring to the same parts as in Fig. 2.

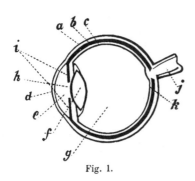

Fig. 1.

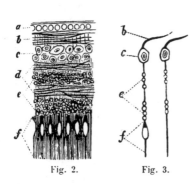

Fig. 2.   Fig. 3.

When a ray of light enters the eye, it passes through the humours or lenses, and is formed by them into an image, on the choroid, of the object looked at. The extremities of the rods and cones have the power of appreciating the image there formed, and convey it up through the ultimate parts of the retina (Fig. 2), thence along the optic nerve fibres to the brain. We are inclined to regard the extremities of the rods and cones as the true seat of perception, in consequence of observing a considerable distance between the retinal blood-vessels and the choroid, when performing Purkinje's experiment* This experiment consists in passing a lighted candle slowly to and fro before the eyes, at about two or three inches from the nose, when the retinal vessels will exhibit themselves before the observer not unlike branching trees. They may be seen by daylight, by passing the large teeth of an ordinary comb slowly backwards and forwards before the eye, whilst looking on a smooth sheet of paper, or upon the sky. Fig. 4 represents those of the left eye, as seen by candlelight. The spot marked k is the exact centre of the retina. (The same letter marks the same spot in Fig. 1.) It is the seat of most distinct vision. j is the entrance of the optic nerve (Figs. 4 and 1), from centre of which the retinal artery will be seen emerging and spreading over the entire retina ; but in the diagram that part only is represented

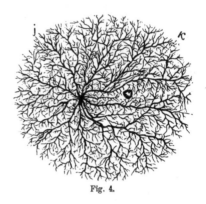

Fig. 4.

which could be seen tolerably distinct. The background to the artery appears of a pale red, except at the part occupied by the optic nerve, where it is white.

After this rapid glance at so complicated a structure, and bearing in mind that some persons can see its several parts with vastly greater facility than others, it cannot be a matter of surprise that individuals not aware of these facts are, now and then—especially at night, and when carrying a light about—startled by what they fancy an apparition, but which is in reality nothing more than some part of the structures above considered. A lady assures us that she saw the ghost of her husband as she was going downstairs with a lighted candle in her hand. The spot k, Fig. 4, when seen

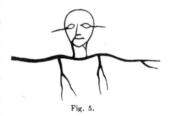

Fig. 5.

against a wall a few feet distant, appears about the size of a human head, and wants very little to furnish it with features. Figured paper on the wall, and a host of other things, may supply them, or even the retinal artery, which often lends body and limbs. (Fig. 5.)

*This distance can easily be perceived by getting an impression on the retina, according to the "Directions," page 4, and then, on performing the above experiment, the arterial ramifications and the central spot will be distinctly perceived to move over the spectral figure.

Besides the above-mentioned structures, there are others which may play an important part in these illusions, especially the common *muscae volantes*, so called from their resemblance to flying flies. They consist of cells and filaments, the *débris* of the structures of the eye, and float about in its humours. That some of them exist very near the retina appears evident from the fact that, on placing the eyes close to a gauze wire blind, distinct miniature images of parts of the gauze will be seen in them. (Fig. 6.)

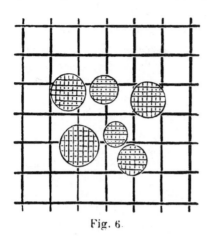

Fig. 6.

We now pass on to consider some of the leading properties of Light. There have been many theories propounded from time to time in order to explain the various phenomena connected with this subject, but only one accords well with all, and that is called the undulatory or vibratory theory, which, from its numerous complications, will compel us to confine ourselves to a consideration of that part only which is necessary to our present use. This theory regards light as the vibrations of an imponderable ether pervading all space, the number of these vibrations varying in a given time for each of the three primary colours—blue, yellow, and red—the greatest number producing blue, the least red, and an intermediate number yellow, all other colours being produced by the combination of these in various proportions. Any two of the three primary colours mixed together makes the complementary colour to the third, and the third is also complementary to it. Thus, blue and yellow make green, which is the complementary colour to red; red and blue make purple, complementary to yellow; yellow and red make orange, complementary to blue. When the three primary colours are mixed together, white is the result: so that when a ray of white light falls upon a piece of paper, and all the vibrations are equally reflected, the paper will appear white, and if they are all absorbed, it will appear black; but, if the paper absorbs some and reflects others, it will appear coloured. Thus, if it absorbs those producing red, it will appear green, from the mixture of the vibrations producing blue and yellow; and if it absorbs blue and yellow, and reflects red, then it will appear red. In this manner any object we look at will appear of any particular colour, according to which vibrations it absorbs and which it reflects.

The retina is so admirably constructed that it is susceptible of different impressions of colour by these different vibrations, except in the case of a few individuals, who are either blind to all colour, and therefore see everything black or white, and their intermediate shades, or who are blind to only one or two colours.

When we look steadily at a red object for a few seconds, that part of the retina on which the image impinges begins to get less sensitive to vibrations producing red, but more sensitive to those producing blue and yellow ; so that on turning the eye away from the red object, and permitting a little white light to enter it, that part of the retina which received the red image will, in consequence of its diminished sensibility to that colour, and its exalted sensibility to blue and yellow, be able to perceive the two latter colours best, and by their mixture will give rise to a green image of the red object. The same thing will be observed with all the other colours; the secondary image or spectre always appearing of the complementary colour to the object from which the impression is obtained.

The duration and vividness of these impressions on the retina vary greatly in different individuals, and can be procured from almost any object. A person may, after looking steadily, and as often happens, unconsciously for a short time at printed or painted figures, on paper, porcelain, &c., see, on turning the head in some other direction, a life-sized or colossal spectre (the spectre appears larger the greater the distance of the surface against which it is seen), and there can be little doubt but that many of the reputed ghosts originate in this manner.

3

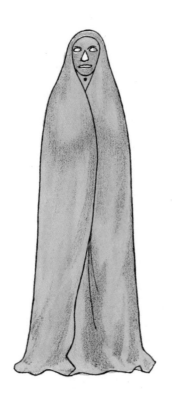

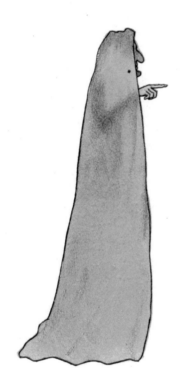

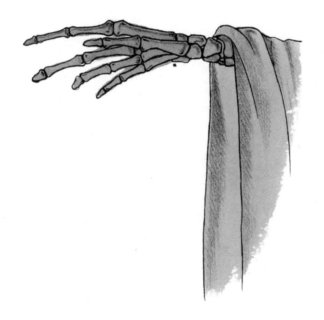

10

11

12

14

15

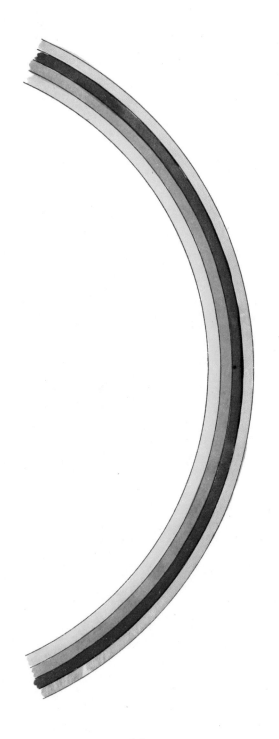

14